trotman

Real Life GUIDES

Dee Pilgrim

2nd edition

Real Life Guides: Carpentry & Cabinet-Making
This second edition published in 2008 by Trotman Publishing, a division of Crimson Publishing Ltd, Westminster House, Kew Road, Richmond, Surrey TW9 2ND

First edition published in 2003 by
Trotman and Co Ltd

Author Dee Pilgrim
Advertising Sarah Talbot, Advertising Sales Director (020 8334 1617)

Design by XAB

British Library Cataloguing in Publications Data
A catalogue record for this book is available from the British Library

ISBN 978 1 84455 153 8

Typeset by Ellipsis Books Ltd, Glasgow
Printed and bound in Great Britain by TJ International Ltd, Padstow, Cornwall

Real Life

GUIDES

CARPENTRY & CABINET-MAKING

REAL LIFE GUIDES

Practical guides for practical people

In this increasingly sophisticated world the need for manually skilled people to build our homes, cut our hair, fix our boilers, and to make our cars go is greater than ever. As things progress, so the level of training and competence required of our skilled manual workers increases.

In this series of career guides from Trotman, we look in detail at what it takes to train for, get into and be successful at a wide spectrum of practical careers. The *Real Life Guides* aim to inform and inspire young people and adults alike by providing comprehensive yet hard-hitting and often blunt information about what it takes to succeed in these careers.

Other titles in the series are:
Real Life Guides: The Armed Forces
Real Life Guides: The Beauty Industry, 2nd edition
Real Life Guides: Care
Real Life Guides: Catering, 2nd edition
Real Life Guides: Construction, 2nd edition
Real Life Guides: Distribution & Logistics
Real Life Guides: Electrician, 2nd edition
Real Life Guides: The Fire Service
Real Life Guides: Hairdressing, 2nd edition
Real Life Guides: The Motor Industry, 2nd edition
Real Life Guides: Passenger Transport
Real Life Guides: Plumbing, 2nd edition
Real Life Guides: The Police Force
Real Life Guides: Retail, 2nd edition
Real Life Guides: Transport
Real Life Guides: Working Outdoors
Real Life Guides: Working with Animals & Wildlife, 2nd edition
Real Life Guides: Working with Young People

Real Life
GUIDES
CONTENTS

About the author

Dee Pilgrim completed the pre-entry, periodical journalism course at the London College of Printing before working on a variety of music and women's titles. As a freelancer and full-time member of staff she has written numerous articles and interviews for *Company*, *Cosmopolitan*, *New Woman*, *Woman's Journal* and *Weight Watcher's* magazines. As a freelancer for Independent Magazines she concentrated on celebrity interviews and film, theatre and restaurant reviews for such titles as *Ms London*, *Girl About Town*, *LAM* and *Nine to Five* magazines, and in her capacity as a critic she has appeared on both radio and television. When not attending film screenings she is active within the Critics' Circle, co-writes songs and is currently engaged in writing the narrative to an as yet unpublished trilogy of children's illustrated books.

Acknowledgements

Thank you to Mark Pollecut of Cheeky Monkey Treehouses, David Pearham of the Building Crafts College, Samantha Jay Compton, Cheryl Mattey and Paul Magerum for their contributions to this book. Many thanks to the Carpenters' Company and staff at the Building Crafts College. For detailed information about training in the industry, a big thank you to ConstructionSkills.

Foreword

If you enjoy working with your hands, have a passion for natural materials and want to make a contribution to the world around you, then why not consider a career in carpentry? Within the construction industry, carpentry is one of the skilled trades most in demand and job opportunities exist across the UK.

To be a successful carpenter you will need to have plenty of practical ability and be prepared to get involved in physical, manual labour. You will need to have a good eye for detail and be able to follow drawings, plans and written or spoken instructions. Have you got what it takes?

Carpenters work on building sites or in workshops making or repairing a variety of structures, such as doors, windows, staircases, floorboards, roof joints and partitions. The skilled input of a carpenter is helping to hold up the ceiling above your head right now! They use traditional woodworking tools as well as specialised power or hand tools to cut, shape, smooth and finish a wide range of different types of wood.

There are lots of different employment opportunities for skilled carpenters. As well as working for a construction company on new builds you could get involved in building maintenance or repair, shopfitting or property development.

You could even become self-employed, choosing the jobs that suit you, or run your own business.

A good way to start you career is as an apprentice carpenter, combining learning on the job with studying at college. To become a fully qualified professional you will need to gain an industry-recognised NVQ. Find out how City & Guilds can help you to achieve the qualifications that you need by visiting www.cityandguilds.com. City & Guilds are delighted to be part of the Trotman *Real Life Guides* series to help raise your awareness of these vocational qualifications.

Introduction

The art of carpentry, or of working with wood, is almost as old as the human race. Ever since man realised he could use branches and timbers to build himself a shelter, he's been evolving tools and methods to turn the trees around him into homes, furniture, wheels, plates, bowls – and even into other tools and weapons such as bows and arrows. Although new, man-made materials such as steel and glass have come along, wood is still used extensively in our homes, the places where we work and even in leisure establishments such as restaurants and bars. That's why the skills of the carpenter will never become obsolete. In fact, carpenters are as much in demand today as they have ever been, which is why training to be a carpenter or cabinet-maker can mean having a job for life.

You may think of a carpenter as a Bob the Builder character, travelling around in his van fitting new windows, cupboards and doors in people's private homes, but this is only one of a myriad of different jobs carpenters undertake. The truth is there are shortages of skilled tradesmen in all areas. In fact, according to a report by ConstructionSkills (formerly the Construction Industry Training Board, the Sector Skills Council for the construction sector) an average of 87,600 new construction workers will need to be recruited and trained each year just to keep up with demand, and woodworking skills is one of the biggest growth areas, with an 11 per cent increase in numbers needed over the next five years (that's approximately 12,500 new recruits each year).

There are many reasons why there is such a shortfall of people with the necessary carpentry skills. Construction-Skills admits that in the past the whole of the construction industry suffered from a bad image and was not the first choice of career for young people, as they felt workers with manual skills were not respected in the community. In particular, the industry struggled to recruit women and young Asian and black people because they were underrepresented in the industry as a whole. Construction-Skills has now addressed this issue with its new, national recruitment campaign, Positive Image. Also, a lot of new trainees moved to other jobs where they could use their manual skills – like the car industry in Birmingham – that paid about three times what carpentry could. This has changed with better pay and conditions, although a lot of trained carpenters still move from the construction site to boat-building, especially canal boats and other small pleasure craft.

Two other factors that have changed are the huge boom in multi-million-pound building projects across the country (most especially the Olympics) and the public's interest in green and environmental issues leading to the desire to see wooden buildings preserved.

People who decide to train to be a carpenter can now get highly paid for their skills. Conservation is one of the biggest growth areas. A lot of money has been pumped into conservation by English Heritage, Heritage Scotland and the Church of England.

It seems everywhere you look these days you can see cranes and scaffolding as new buildings go up. And it isn't only huge projects like the development of the Olympic Village or St David's 2 Centre in Cardiff that are being built. After years of underinvestment, when not enough domestic homes were being built to keep up with demand, there is now a drive on to build 200,000 new homes a year. In the South-East alone, over the last three years, the government set aside three-quarters of a billion pounds for building affordable housing. That figure will rise to more than £20 billion over the next 12 years. In fact, it's been estimated that a staggering 1.5 million new homes will have been built by 2016 and that's an awful lot of houses for carpenters to work on.

While these new buildings are going up, Britain's older, or 'heritage', buildings need constant renovation. Not only have English Heritage, Heritage Scotland and the Church of England pumped a lot of money into the conservation of buildings such as Tudor timber-framed houses, but also more ordinary dwellings need the carpenter's attention. David Pearham, the Deputy Director and Chief Instructor at the Building Crafts College (BCC) in Stratford in East London, explains that conservation is one of the biggest growth areas. Indeed, many of the BCC's former students are currently working on the rebuild of the famous sailing vessel the *Cutty Sark*, in Greenwich. 'If you take the whole of the construction industry, 48 per cent of construction work is what we would class as renovation (making good buildings that already exist),' says Pearham. 'Of that, 24 per cent is working on buildings built before 1911, so already a quarter of construction work is within the heritage industry. That's where most of our trainees should be looking to be employed: repairing box-frame sash windows, working with architects to restore ornate cornices, or replacing skirting boards. People who decide to train to be a carpenter can now get highly paid for their skills.'

Another area where there has been great expansion is jobs for local freelance carpenters. The massive rise in interest in DIY and house decoration has actually done carpenters a favour and has certainly raised the carpenter's profile. TV shows such as *Grand Designs* in particular have shown ordinary members of the public that modern doesn't have to mean glass or plastic but can mean a beautiful piece of wood. New wood floors, or a beautifully turned balustrade for the staircase, even building a full green oak timber-frame house, are some of the projects for which you need a well-trained carpenter. The interest in interior design has also re-sparked an interest in the skills of the cabinet-maker, with increasing demand for individual, handmade pieces of furniture. 'There's a better understanding from the general public about quality items, and bespoke cabinet-makers and joiners are very busy at the moment because people are now prepared to spend money on quality,' says Pearham. 'They've had enough of mass-produced items from the likes of MFI and Ikea and are prepared to wait and save for fitted furniture.'

In recent years, as more countries have joined the European Union, there has been an influx of Eastern European builders and carpenters into the country (especially from Poland,

Estonia and Albania) but it seems with all the construction going on there is still more than enough work to go round. According to the most recent records available from the Office of National Statistics (ONS), 289,916 people classed themselves as either part-time or full-time carpenters in June 2007. As experienced carpenters retire or move up the career ladder into management positions, there is (and will continue to be) an ongoing need for young trainees to come through the system. 'There is a real market for rewarding, quality work for wood occupationalists,' says Pearham. 'Making things is a very rewarding way of life. Also, because carpenters are involved in all aspects of construction it develops a mindset for problem-solving, so it's no surprise 95 per cent of managers in the industry come from a carpentry background.'

It seems wood, in all its glorious colours and textures, is making a comeback. All of this means it has never been a better time to train in carpentry, for although the work is highly skilled and demanding, the rewards – both personal and financial – can be immense.

DID YOU KNOW?

Carpenters who make wooden guitars by hand are known as luthiers.

This book is intended to help you decide whether a career in carpentry is for you. It will explain what the various branches of carpentry actually entail and what skills you will need to enable you to do the job. It will offer information on what qualifications you will need to progress to a vocational course in carpentry and also what the broader opportunities in carpentry are. Whether you want to work as one of a team, or for yourself and by yourself, working with wood to produce something practical, such as a roof beam, or something beautiful, such as an exquisite marquetry table, can be a wonderfully fulfilling way of earning your living. And remember, there's somebody out there right now crying out for a good carpenter!

MARK POLLECUT

Success story

THE TREE HOUSE BUILDER

Mark Pollecut is the proprietor, designer and construction engineer for Cheeky Monkey Tree Houses, a company he set up five years ago. 46-year-old Mark is actually an engineer by trade. He took a BTEC HNC in Engineering at Southall College and continued his training with the British Airport Authority (BAA). There, he worked in the drawing office and in different departments at various airports. However, after he and his wife started a family he became a househusband and it was when he built his children a tree house that his new career took off. He started to build tree houses for his children's friends at school and that led him to the idea of having a small business making tree houses. He enjoyed the work so much he stepped up production and 2007 was his busiest year yet and he has had to take on another part-time member of staff to keep up with demand. Cheeky Monkey Tree Houses is based in Surrey, but Mark builds tree houses all over Britain and Europe and has a lot of enquiries from America.

Apart from being skilled with carpentry tools you also need to have common sense and good child awareness because often the client's children are around during the build, so you can't leave tools lying around.

'I design the tree houses, I oversee the builds and I am part of the builds as well – yes I do get my hands dirty, although I'm trying to get away from that as I need to concentrate on the managerial side and the actual designing. Our tree houses are constructed entirely of wood – they are wooden framed and the cladding is like wooden decking material — and they look like Swiss log cabins. They are mostly self-supporting, so they don't have to rely on the tree for support, in fact we try to minimise contact with the tree. Most of the cabins are wrapped around the tree, which allows people to have the size they want, and they can sleep in them because they are fully insulated and chunky. They are very sturdy and built to last and have different levels of decking and towers – the biggest I have made so far is a seven berth.

Our tree houses are constructed entirely of wood – they are wooden framed and the cladding is like wooden decking material – and they look like Swiss log cabins.

'I never make two tree houses the same. I like to do my drawings freehand on a drawing board so the client can see the design as it is being done. I encourage a lot of input from my clients, especially as to where they want the windows and doors to go. In fact, the thing I like the most is the designing and the drawing because I have a certain talent for it, and I love working things out on the drawing board. I enjoy meeting the clients and I enjoy working with my hands. I also like meeting all the other people involved and the freedom this business gives me. It is wonderful to experience the excitement people feel when they see their completed tree houses. Luckily, the way we put the structures together, half built in a workshop and then erected and finished on site, it doesn't take long to construct the waterproof shell, so we are rarely interrupted by the rain or the cold.

'The other thing I really like is the fact I am working with wood, because it really lets you use your imagination because there is so much you can do with it. At the moment I'm working with just one other, self-employed person, but I really need to employ some more carpenters. Apart from being skilled with carpentry tools, especially machine tools, you also need to have common sense and good child awareness because often the client's children are around during the build so you can't leave tools lying around. You need to be adaptable and have fresh ideas all the time. I think you need to be friendly and trustworthy and above all, to be passionate about it.

DID YOU KNOW?

Celebrity actor Harrison Ford actually worked as a carpenter in Los Angeles before hitting the big time. His carpentry skills stood him in good stead when he was working on the film Witness, where in one scene he helped erect a timber frame house.

I really like the fact I am working with wood, because it really lets you use your imagination because there is so much you can do with it.

'I'd like to see the business progressing, getting bigger slowly and at a sustainable rate. At the moment we are building one tree house a month and I would like to up production to two a month, but that would mean employing more people.'

For more information visit www.cheekymonkeytreehouses.co.uk.

2

What's the story?

If your parents have ever had new cupboards built at home, or your school has had wooden window frames replaced, you will have seen a carpenter at work first hand. However, the maintenance worker or freelancer who travels around building or fixing wooden structures such as kitchen cabinets, staircases and box sash windows is only the tip of the iceberg: there's much more to the carpentry story than that. Carpenters are everywhere, even when you are not aware of their activities, and they need to be highly trained and skilled in order to do their jobs. But what does being a carpenter or a joiner mean? What about a cabinet-maker? Basically, carpentry is the art of cutting, joining and framing together timbers essential to the stability of a structure, while joinery is the art of dressing, framing, joining and fixing wood for the finishings of houses. So, in basic carpentry, it doesn't really matter what it looks like as long as it does its job, while joinery not only has to do its job, it also has to look good because it will be on view. A cabinet-maker either puts together ready-cut parts of furniture in a workshop, or is a skilled joiner producing hand made desks, chests of drawers, and other pieces of furniture of high quality.

Cabinet-makers sometimes work with bespoke furniture designers who often collaborate with their customers to produce one-off pieces of furniture. Designers make drawings of their own original ideas or work to a brief given to them by their employer or client. Some are employed by large companies to create designs for mass production (the kind of kitchen or bedroom units you see in furniture warehouses such as Ikea), others are freelance, but all designers will work closely with the people who will actually be producing the finished piece of furniture in order to make sure it is fit for purpose (functional) and looks good. Other people involved in wood occupations include those who repair or restore furniture. These include furniture polishers (including French polishers), who restore scratched or stained furniture with specialist cleaning fluids, waxes and polishes, and upholsterers, who re-cover pieces of furniture when the fabric wears out or gets dirty. Many of these tend of be independent workers who have set up their own specialist businesses. Meanwhile wood-turners produce bowls and other beautiful objects by whittling away wood that is turned on a lathe, and picture framers select the wood and finish of a frame to suit the picture or mirror it contains. Finally, specialist timber-frame house builders, boat builders, and even wooden tree house designers, like Mark Pollecut in the previous chapter, are out there working with wood in ways you may never have imagined. These jobs all involve a certain amount of interaction with clients and with suppliers and so good social skills are a must. They also demand a good head for business, because in the majority of instances the independent worker is his or her own boss.

If setting up your own business fills you with foreboding, there are still plenty of other areas in carpentry where you can work as an employee. Nearly every building you see around you, even those that are predominantly constructed from concrete, metal and glass, will have required the services of a carpenter at some point while being built. Here are the most common places you will find carpenters, joiners and cabinet-makers at work, with a brief description of what their different titles and specialities are.

CONSTRUCTION SITES

This is where **formworkers** make temporary structures that are used like moulds to support wet concrete before it sets, **joiners** erect and fix floor joists and roof timbers, **first fix carpenters** finish off the internal woodwork, such as partition walls, and **second fix carpenters** do the final bits, such as skirting boards, door surrounds and fitted wardrobes. Carpenters on construction sites work with power tools and are surrounded by a lot of heavy machinery, so they must be very safety conscious. The work on site can be very physically demanding.

WORKSHOPS

The workshop is where **bench joiners** construct fitted furniture, panelled doors, windows, and staircases, and where the **cabinet-maker** makes anything from desks to chests of drawers. This is also where independent **picture-framers** usually work (the workshop may also have a shopfront where customers can order the frames), while specialist **polishers** and **restorers** can work either in a workshop or, if it's a very heavy or fragile item, wherever the particular piece of furniture happens to be.

MACHINE SHOPS

Machinists prepare and shape rough timbers into floorboards, skirting boards, and panels using specialist equipment.

TV/FILM/THEATRE SETS

When scenery has to be built for a TV production or a film it may be constructed in sections in a specialist workshop before being erected on a stage or in a studio. Alternatively, it may be built in the studio itself.

Once again, the work in all these locations can involve the lifting and carrying of heavy beams or pre-formed panels and workers must keep an eye on health and safety regulations.

ON SITE

Freelance carpenters and **shopfitters** (who specialise in producing the fronts and interiors for restaurants, banks, and shops) may do most of their work on site ie in people's homes or in retail outlets.

DESIGN STUDIOS

Furniture designers, if employed by big companies, usually work in design studios and consultancy practices. Here, they produce drawings and can build mock-ups of their designs to see if they work. Self-employed designers like Mark Pollecut may work from home or from rented studios.

Obviously, carpenters who want to do the more intricate, skilled jobs need a lot more training and experience than those whose duties are quite basic. However, there is no clear route of progression through carpentry: this will very much depend on your own skills and interests. For example, you may decide to become a **construction manager**, responsible for keeping a job on budget and meeting deadlines while overseeing a construction crew, and for this you will need a university degree or many years of experience in the business. The degree route (or equivalent level higher education qualification) is also the usual one for furniture designers.

If you like the security of working regular hours for someone else, you may decide to become a **maintenance carpenter** working for a local authority, for example, where your duties will involve the upkeep of public buildings such as schools or council offices. For this career, vocational qualifications are more appropriate and there are a number of ways you can train.

The traditional way to train as a carpenter has always been to become an apprentice, working for a company or individual while learning on the job. This has benefits for the company, in that it gets an extra pair of hands to help out with the more menial tasks, and for the apprentice, who receives a wage, or at least expenses, while learning. Learning while working is still the favoured method of training for carpenters and joiners. The Training chapter (page 43) looks at the training and qualifications available in more detail.

DID YOU KNOW?

Most of the wood used in construction in Britain comes from European forests. Due to intensive planting, these increased in size by 30 per cent between 1990 and 2000 and are still expanding.

THE WONDERFUL WORLD OF WOOD

One of the first things any trainee carpenter learns is how to differentiate between woods from various trees and what jobs/functions they are best suited for. In the past, hardwoods such as oak were routinely used for house and boat construction. However, most hardwoods grow slowly and many are now in very short supply and are consequently very expensive to use. In fact, some hardwoods such as teak and mahogany are now endangered species. In contrast, softwoods, such as pine, grow quickly and are quite plentiful so they are much more affordable. Below is a list of the major woods and their uses. The Forest Stewardship Council (FSC) is a globally recognised scheme whereby only wood from sustainable sources is marked with the FSC logo, meaning you can instantly tell if it has been ethically sourced. For more information go to www.fsc.org.

- **Ash:** Hardwood. Light-coloured furniture and panelling, tool handles.
- **Beech:** Hardwood. Furniture, woodblock flooring, musical instruments.
- **Cedar:** Hardwood. Cabinet-making, boat building (stocks of cedar are endangered.)
- **Ebony:** Hardwood. Cutlery work, doorknobs. (Stocks of ebony are endangered.)
- **Elm:** Hardwood. External cladding, floorboards.
- **Douglas fir:** Softwood. Plywood, structural work.
- **Larch:** Softwood. General-purpose timber.
- **Mahogany:** Hardwood. High-quality furniture. (Stocks of mahogany are endangered.)
- **Oak:** Hardwood. Joinery, fittings, furniture, flooring, timber beams.
- **Pine:** Softwood. General utility work, plywood.
- **Spruce:** Softwood. General utility work.
- **Teak:** Hardwood. High-quality furniture, flooring. (Stocks of teak are endangered.)
- **Walnut:** Hardwood. Decorative panelling. (Stocks of walnut are very vulnerable.)

THE SHOP FITTERS: A EDMONDS & CO LTD

This highly renowned shop-fitting company was founded in Birmingham back in 1870 by one John Mason. It was employed by the shopkeepers in and around Birmingham to fit out their mainly wooden shops. In 1904 Arthur Edmonds was made the General Manager and the company expanded from solely concentrating on shop-front design and fabrication to also making show cases for museums and doing the interior fittings for banks. In 1908 the company changed its name to A Edmonds & Co Ltd and during the First World War it produced Lewis gun chests and also mahogany aircraft propellers for Daimler.

During the Second World War the company premises were almost destroyed by a bomb and it was the A Edmonds workforce that set about making the repairs. It now has premises in both Birmingham and London and has 120 employees with a turnover of £8.6 million in 2006. It can produce individually designed installations with hardwood joinery and undertakes both new construction and refurbishment contracts.

Some of its major projects include:

- Production and fitting of the oak panelling, stairs and doorsets at Birmingham University Great Hall and Senate Chamber
- The counters, curved lecture pods and veneered panelling at Aston University Business School

- The internal joinery elements of Cardiff Millennium Centre
- High-quality boardroom tables for Blackstone, London.

In 2006 the company won the prestigious National Association of Shopfitters Design Partnership Award for its work on the extension of the Grade I-listed Senior Common Room at St John's College in Oxford. Working in conjunction with the architect MacCormac Jamieson Pritchard, A Edmonson helped fit veneered panelling in keeping with the older areas of the building, and produced bespoke dining chairs and tables. Commenting on the furniture the judges said 'the bespoke chairs are simple, elegant and beautifully made, but also incredibly comfortable.'

Sources: A Edmonson & Co Ltd/The National Association of Shopfitters.

3

Tools of the trade

When people think of the tools used in carpentry they usually visualise a workroom stocked with saws, drills, hammers and chisels. Of course, these are all vital in woodworking and a carpenter could not get by without them. Yet there are other, more personal, tools that are just as important. While hard work and determination are needed in order to progress in any job, there are more specific skills and strengths that a carpenter will rely on throughout his or her working life. By pinpointing these skills you will be able to see whether you really do have what it takes to make a career in carpentry. If you believe carpentry is for you, you can purchase hammers and saws as you train and work, but for now, see how many of these personal tools you already possess.

- A carpenter does not spend their working hours sitting behind a desk – they are more likely to be on their feet making it! This will involve picking out timbers, bringing them to the workshop, sawing them, joining them and finishing them. All this hard physical work can be pretty demanding, so being physically strong and having stamina are essential for a job in carpentry. **Strength and physical fitness** are even more important for carpenters who work on building sites, erecting roof timbers or wooden frameworks for houses. According to David Pearham, carpenters working on sites tend to switch to management jobs as they get older and the physical aspect of the work becomes more difficult for them, whereas it is not unknown for bench joiners to carry on until they are 65. Needing physical strength is one reason why women sometimes feel they can't become a carpenter. However, Cheryl Mattey, the Bench Joinery Instructor at Building Crafts College, worked as a freelance carpenter for many years before becoming a teacher, and says the techniques carpenters are taught to lift and carry wood mean that anyone who enjoys normal physical health can do the job.

- If you decide to become a carpenter working on the construction of buildings, you really do need to have a **head for heights**. You could well find yourself on the top of a two-storey building fitting roof timbers, or up a ladder fixing the wooden panelling on a gable end. Bear in mind, you might be doing this in all weather conditions.
- Although you will be taught the practical skills of carpentry during your training, it does help if you are already quite deft. Good **hand-eye co-ordination and dexterity** will certainly make your life easier, especially if you decide to specialise in the more intricate sides of carpentry such as finishing, cabinet-making or marquetry. Doing some kind of work experience or a course that is hands on (metalwork, say) will let you see just how dextrous you really are.
- An **ability to draw** and having a **creative eye** are two of the most important skills needed by the furniture designer. Although it is true many designers use CAD (computer-aided design) these days, the ability to pull out a sheet of paper and make a quick sketch to show a client what you are thinking of designing is a wonderfully helpful skill to have. In fact, for many designers the actual drawing – being able to put the idea they have in their head down on paper – is one of the most satisfying parts of the job. If you are already good at art then design could be where your future career lies.
- It's not just your hands that have to be deft: you'll be using your brain quite a lot to work out how much wood you need for a certain job, what angle a joint should be and the correct

thickness of certain timbers and panels. This means you have to be good at mathematics and geometry, and extremely precise with your measurements. You also need to be able to read and correctly interpret construction pictures and diagrams. It's no good building a lovely wooden door only to find it is too big to fit in the doorframe. **Accuracy** is essential.

- Anywhere that construction is taking place and power tools are being used can be dangerous. A good carpenter is always conscious of the **health and safety** aspects of each individual job. In carpentry hazards lurk everywhere. Building timbers are heavy, you'll often be working high up on ladders or scaffolding, circular saws are very sharp, and electric power tools must be treated with the utmost respect. This is particularly true on building sites where heavy plant is also in use, so being constantly aware, cautious and careful are great skills to have.
- Being able to understand what other people are talking about and making yourself understood in return are essential to carpenters. If you are working as part of a team you need to know who is doing what and when, especially as you will be liaising with other skilled workers such as plumbers and electricians. (Pipes and wires will have to be installed behind that lovely wooden wall panelling.) Even freelance carpenters who work alone need to have good **communication skills** as they will be ordering raw materials from suppliers and talking to their clients about their needs and requirements. Instructions need to be clear and concise so that everyone knows exactly what is going on. In fact, having these 'soft' skills – being a good listener and also being able to make yourself understood easily – is often why some self-employed carpenters keep getting repeat work and good 'word of mouth'. If a client has had a good experience with one carpenter, they are more likely to recommend that carpenter to friends and neighbours.
- The same goes for having **tact and diplomacy**. If a delivery fails to arrive, or a customer is unhappy with a job it is no use

screaming and shouting about it. You have to cooperate with other people and if they can see you dealing with difficulties politely and efficiently they are much more likely to use you again, and to recommend you to other people.

- **Punctuality** is vital wherever you are working. If deliveries are being made to the workshop, you have to be there to make sure they are actually what you ordered. If you tell a client you will be arriving on site at 9am sharp, they will not be amused if you turn up at midday. People lead busy lives and haven't got the time to sit around waiting for you to arrive when you feel like it. Being punctual is a sign of professionalism, so use it to your advantage. This is, maybe surprisingly, particularly true on building sites. The reason for this is builders work longer hours in summer to make maximum use of the hours of sunlight. Because of this work on the site can start as early as 6am, and so turning up at 9am means you've wasted three hours of the contractor's time.

We've looked at the skills you need in order to get on: now let's look at certain physical conditions that could hold you back, and aspects of working in carpentry that may not suit you personally. This does not mean you could not work anywhere within the sector. It may be certain other areas of the industry would be more suitable for you – such as working in a more office-based environment. Have a good think about the following conditions before making up your mind that this is the career for you.

- If you suffer from **vertigo**, a building site will most certainly not be for you: carpenters are required to climb up scaffolding and ladders in order to fit timbers on roofs. Just think of the carpenters who have to fix church spires!
- Working with wood creates a lot of dust, and even though safety standards and measures have improved dramatically in recent years, if you suffer from **breathing problems** or

have allergies to dust this could cause difficulties. No matter how good your face mask or the ventilation in the workshop, it is inevitable that some wood dust particles will remain in the air.

- Many different adhesives are used in carpentry, for gluing joints, and for fixing wood veneer and laminates. If you suffer from **skin sensitivities** these can cause allergic reactions such as contact dermatitis.
- If you are **clumsy** then carpentry is probably not for you. Getting splinters is one of the acknowledged hazards of the trade and you should expect them, but you don't want to lose the top of your thumb to a circular saw, hit your hand with a hammer or damage yourself with any of the other tools a carpenter uses every day. Bear in mind that your clumsiness could also cause accidents to other people.
- Many carpenters find they are forced to change job due to **back problems**. As mentioned previously, the correct and least damaging ways to lift and carry wood are taught in college, but anyone with a family history of back trouble should be aware that bad backs are common in the industry and are one of the reasons why many carpenters switch over to management or overseeing positions after a certain amount of time in the industry.
- If you can't add up to save your life or think geometry is an alien language, you won't be able to cope with the complicated equations necessary to produce a curved wooden structure such as an arched doorway or garden bridge. Mathematics is part of the job, so **if you hate maths** this really isn't for you.
- Many carpenters still work in the traditional way, 'following' the work around the country. This means they **travel** all over Great Britain and abroad, working on various building projects as they arise. If you don't want to spend long periods of time away from home, this particular type of carpentry work will not suit you.

You should now have a better idea of some of the skills and strengths needed for the job and some of the downsides to the industry. If you are still enthusiastic about carpentry, complete the short **quiz** below to discover what your general knowledge of the work is like. All you have to do is choose the multiple-choice answer you believe is correct in each case. This is really just a fun way of seeing if you know as much about carpentry as you think you do, so don't worry if you get some wrong. Answers appear at the end of the quiz.

1. Which of the following tools would you NOT expect to find in a carpenter's workshop?

   A. Hammer
   B. Chisel
   C. Rotary whisk.

2. You are building a bespoke kitchen for a client and he gives you a sample of the wood he wants you to use to make his work surfaces. You approach your usual supplier who has the right quantities of that particular wood in stock, but it is a totally different colour from the sample. Do you:

   A. Go ahead and order the timber anyway? Well, it's what the client said they wanted.
   B. Go back to the client and make sure this is the timber they want you to use?
   C. Contact other timber merchants in the area to see if they have the correct coloured wood in stock?

3. What is the correct amount of wastage you should estimate into any job?

   A. None, you are sure of your measurements
   B. 10 per cent
   C. 50 per cent.

4. You have produced an estimate for a client on quite a big and costly job because you are using expensive oak timber. However, after you've done the estimate and before you can purchase the timber the price of oak suddenly rockets. Do you:

   A. Immediately contact the client and explain what has happened?
   B. Go ahead with the job?
   C. Get on the internet and see if anyone has stocks of oak at lower prices?

5. All carpenters work to a degree of tolerance (the amount their measurements can be 'out'). What is considered to be the maximum amount of tolerance for a trained carpenter?

   A. 5mm
   B. 0.5mm
   C. 3mm

6. You are about to use a power tool on site. It runs through a transformer, but what is the correct voltage?

   A. 50 volts
   B. 110 volts
   C. 420 volts.

7. Which of the following is not standard safety gear for a carpenter?

   A. Steel-toe-capped shoes
   B. Facemask
   C. Ear defenders.

8. You have a job in a private house where the client is a keen DIYer. He keeps telling you he would not do things

the way you are doing them and seems to want you to carry out the job his way. Do you:

A. Explain diplomatically that you are a trained professional and you know exactly what you are doing?
B. Do it his way just to shut him up?
C. Agree to do it his way but say that will add extra costs to the job and so you will have to increase the bill?

9. What is a cornice?

A. Something you are served ice cream in.
B. It's a particular sort of work surface for kitchens.
C. It's an ornamental moulding.

10. What does MDF stand for?

A. It's the name of a furniture store like DFS.
B. Medium-density fibreboard.
C. Modern design frame-house.

Answers

1. C. You would find a rotary whisk in a kitchen, not in a carpenter's workshop.

2. B. You really need to talk to the client before you go any further. Wood, being a natural material, does vary in colour, but if the wood your timber merchant has in stock is the wrong shade it may not suit the overall design of your client's kitchen. Then again, they may prefer this colour and ask you to go ahead with it. If not, you are going to have to find a supplier who does stock this particular shade of wood, or get your client to choose an alternative.

3. B. You should always factor in a degree of wastage. This is because there may be imperfections in the wood and the shape of whatever you are making may produce offcuts that are of no use to you. Ten per cent is the industry standard.

4. A. This is another occasion when you really need to talk to the client immediately because they may not have the budget to afford the more expensive oak. If they wish to proceed with oak, then you can start looking for other, cheaper suppliers; otherwise, you'll have to look for an alternative timber.

5. B. On most carpentry training courses students are expected to start working within 5mm of tolerance. However, by the end of the course their accuracy should have improved enough for them to work to a 0.5mm tolerance. If you are not sure just how fine that is, take a look on a ruler – it's about the thickness of your thumbnail!

6. B. There is a reason why power tools work off 110 volts. If you were to get an electric shock of 110 volts it would be rather uncomfortable, but it would not kill you. However, 420 volts most certainly would!

7. This is a trick question: all the answers are correct.

8. A. Always remember, you are a professional who has trained to do this job. Explain to the client that you are doing things the way you are because this is the correct method and it will produce the best results. Never compromise your standards.

9. C. A cornice is indeed a type of moulding. It can be a moulded projection on top of a building and it can also take

DID YOU KNOW?

Woodturning differs from most other forms of woodworking in that the wood is turned on a lathe while a stationary tool is used to cut and shape it. Many intricate shapes and designs can be made by turning wood, including chair legs and bowls.

the form of a frieze running round the wall of a room just below the ceiling.

10. B. If you've ever watched the TV show *60 Minute Makeover* you'll know the designers are particularly keen on MDF as it is a very versatile and much cheaper alternative to wood. As a carpenter you may well use MDF frequently on interior jobs.

TOP 10 TOOLS

Above are listed the personal skills or 'tools' you need to get on in carpentry, but here we look at the actual tools a carpenter will work with on a day-to-day basis. You may be surprised by some of them, but these are the carpenter's stock in trade.

1. Hammers

Without a hammer a carpenter really cannot operate. Most professionals like to use a claw hammer – this has a claw opposite the 'hitting' head, with which to prise nails and tacks out of timber, thus making it two tools in one.

2. Retractable tape measure

As you will see as you go through this book, accuracy is all to the professional carpenter. You need to measure everything, as mistakes can be costly. A good-quality tape measure (marked in both metric and imperial units) is a must and the retractable kind is less cumbersome and far more versatile than a ruler.

3. Saws

Carpenters use all kinds of saws, including manual saws and powered saws. The circular saw is a handheld power tool for jobs where straight lines need to be cut, such as cutting timbers in half. Meanwhile, the jigsaw is used to cut curved patterns. In the workshop carpenters also use larger saws such as the table saw (which, as the name suggests, is stationary and fitted to a table) and the compound mitre saw (which can be set to cut various different angles in timber).

4. **Spirit level**

This may be one of the tools you haven't considered, but when you suddenly realise that set of shelves you've just put up is anything but level, you'll be wishing you had invested in one! A spirit level tells you if something is perfectly horizontal or vertical – absolutely essential measurements for everything from building shelves to window frames.

5. **Drills**

Most carpenters now tend to use power drills and the ordless power drill has become very popular (there's no danger of tripping over the cord). However, more heavy-duty jobs will need the added power of the corded drill – great for drilling through masonry. They all come with different sized drill bits depending on the size of hole you require and the nature of the fabric you are drilling through.

6. **Screwdrivers**

Again, you can now get powered (corded and cordless) screwdrivers as well as manual screwdrivers. Because there are various different types of screws out there (see below) you will probably need two screwdriver sets with different sized heads.

7. **Nails and screws**

Much of what carpenters do involves fixing one bit of wood to another and this is often done with nails, tacks and screws. Nails come in all sorts of sizes and with different sorts of heads (from flat to the domed heads of 'hob' nails). The two major types of screws used in Britain are flat head (with a single notch across the centre of the head) and Phillips (with a cross in the centre of the head in which the appropriate Phillips screwdriver fits). Carpenters usually have a wide variety of screws and nails in their toolkits.

8. **Chisels**

This is the tool a carpenter uses for cleaning out material from joints and mortices (a hole in a wooden framework). A good chisel has to be very sharp and so they should always be handled with care. There are varying sizes of chisel, depending on the size of hole required.

9. **Planers**

Carpenters often work with raw, or only roughly finished wood, and a planer is used to shave off very fine layers of the wood to clean it up and to give it a smooth finish. Wood may be sanded and then polished or painted after it has been planed.

10. **Router**

When you want to hollow out a piece of wood but not cut all the way through it, you can't use a saw. This is when a router comes into its own, because it can be used to make indentations and grooves in a piece of wood – either for decoration, or to accommodate a peg or another piece of wood. This is of particular use to the cabinet-maker.

4

DARREN MINETT

Case study 1

THE FREELANCE JOINER

Darren Minett, who is 24 years old, has been a freelance joiner for six years. Before that he was apprenticed to Barratt Homes in the Midlands for four years and studied up to NVQ3. He now works for a small joinery firm that has its own workshop and its own building arm.

'My day starts at 8am when I drive myself in my van to the workshop. When I get there the firm tells me which job I will be on that day – it could be that I will be making something in the workshop or that I will be out working on site. Today I and the other joiner in the firm have been working on a fitted kitchen for a private client. Yesterday we fitted all the units and wall units and today we've gone in and mitred the work-tops and fitted all the cornicing and things like the light rail. We also fitted all the drawers and all of the doors. This really was a two-man job because some of the units were too big to handle alone.

Every day is different. I could never work in a factor. Going to the same place every day would really bore me.

'We are meant to have our first break at half past nine, with lunch at half twelve, but we don't always stop to have breaks if there is something we really need to get done. There is also another break at quarter past three, before we knock off at five, but if we have to work through we get paid overtime or can have time off in lieu.

I have to do all the measuring up and work out what materials I will need, then I go back to the workshop and see if we have the timber in stock.

'If I'm building something in the workshop we have all of our own machinery for cutting it and planing it down such as band saws and cross cut. I have made everything from doors, airing cupboards and gates to flights of stairs. When I was an apprentice Barratt's bought me a small bag of tools and I bought a few cordless tools as I trained. Then, when I went self-employed I went out and bought all my power tools and a bigger van so I can transport everything I need from the workshop to the site.

'If I'm out on site, then the first thing I have to do is go and have a look. I have to do all the measuring up and work out what materials I will need, then I go back to the workshop and see if we have the timber in stock, work out how long the job is going to take us, and then do a cost estimate. If the client agrees to that, then we book a date in for when the job can be done.

'If I am working on a building site the job I like to do the best is roofing. Heights don't bother me and I like it because it is outside and you can really see the job appear. First there's nothing, then there it is, a roof. Not many joiners can do proper, traditional roofing these days and it is hard physical work. The biggest roof job I have ever done was putting in a set of oak roof joists. The timbers were twelve-by-fours and each timber was five metres long. It took five of us to lift each one, they were so heavy!

'At the moment I am spending every Wednesday at college doing an Advanced Vocational Certificate of Education (AVCE). This is a kind of gap two-year course filling in between doing my NVQs and going on to do an HNC in business studies. This is because in the long run I would like to have my own business. The course is really hard because it covers things like town planning and building regulations as well as all the other aspects of the business.

'The other days of the week, I might nip back to the workshop once I've finished just to sort out materials for the next day's job. That's what I love about what I do, there is so much variety and so much freedom. Every day is different. I could never work in a factory. Going to the same place every day would really bore me.'

5

FAQs

By now you will be aware that there's much more to carpentry than there first appears to be. Working with your hands and actually creating things can give you a powerful sense of achievement, but what else will a career in carpentry bring to your life? What are the benefits, financially, socially and personally? This section consists of a number of the most commonly asked questions about getting a job in the industry and will help you to make up your mind if this is really the career path for you.

ONCE QUALIFIED, CAN I MOVE UP THE PROMOTION LADDER QUITE QUICKLY?

That depends on what you decide to do in carpentry. Once your basic training is finished, you will have to decide what area of carpentry you want to specialise in. You could join a bespoke cabinet-makers or a shopfitting firm. You could decide to go into the construction side, working on site. Alternatively, you could decide to go it alone. Traditionally, after their apprenticeship has finished, newly trained carpenters are encouraged by their firms to leave and seek employment with someone else in order to gain more experience in a different work environment. This means you can move up the promotion ladder quickly as you master new skills. However, this is a trade in which you never stop learning, so a quick promotion may not always be best. Sometimes it is better to stay in a job and learn absolutely all you can before moving on. Do bear in mind the fact that as you age you may not be as capable (or as willing) to do the hard, physical work undertaken by carpenters on building sites, so a move into a workshop or into maintenance carpentry when you are older may suit you better. Also, 95 per cent of people in management throughout the construction industry come from a carpentry background and so, even if you love being hands on, this could always be another option for you.

WILL I WORK NINE TO FIVE?

Again, that will depend on which branch of carpentry you decide to pursue. Although some carpenters and cabinet-makers, especially those working in factories, do work a normal 39-hour week from Monday to Friday, many do not. For example, carpenters working on building sites will want to make use of all the hours of daylight they can. This means in the summer they will start earlier and finish later than normal and may work weekends as well. Many carpenters are working to a deadline and might have to work overtime to get the job finished. The freelance carpenter working in people's homes may have to tailor their hours to suit their clients.

WILL I GET TIME OFF FOR HOLIDAYS?

Yes, you will. Once again, the amount of holiday you get will depend on which branch of carpentry you are in, on your employer, and on whether you are self-employed or not. If you are self-employed you will have to calculate how much time you can take off, not only in financial terms, but also in terms of how many job opportunities you are going to miss. If you are employed on site it is normal to dovetail holidays between completing one job and starting the next. Again, because of the need to maximise working hours during the lighter summer months, you may have to take your holidays in autumn or winter.

HOW MUCH CAN I EXPECT TO EARN?

Pay will depend on your experience and expertise. Traditionally, apprentices are not particularly well paid, but this situation is improving. David Pearham of BCC points out that bench joiners tend to earn less than site carpenters because of the nature of the work. 'Bench joiners can earn between £400 and £450 a week,' he says, 'while site carpenters get paid between £600 and £750 a week because the work is more physically demanding and has more of a risk factor.' At the time of writing trainee carpenters and joiners start at around £12,500 per year and as they gain their qualifications this rises to around £14,500, while fully-qualified carpenters and joiners receive £21,000 per year upwards. Cabinet-makers start on around £10,500 rising to £25,000, while wood machinists start on £11,500 rising to £21,000 with qualifications and experience. However, these figures are only intended as guidelines. Furniture designers can expect to start on between £14,000 and £18,000, rising to £25,000, but many earn much more than this. In fact, David Pearham says he personally knows five millionaires who started as carpenters and built up their own businesses, employing other members of staff. Many woodworkers increase their earnings by working overtime or by claiming travel allowances. If they are working in London they may also be paid London weighting.

WILL I HAVE ANY COSTS WHILE I TRAIN OR WHEN I AM FULLY QUALIFIED?

Yes, you will. Although colleges and most firms will supply

you with safety gear such as goggles or face masks while you train, you will probably have to buy your own steel-toed safety boots. However, these are a great investment as you'll most certainly need them once you are qualified. You may also have to contribute to the cost of any expensive hardwoods you use at college – woods such as oak or beech are very costly compared to pine and if you are making a hardwood cabinet or shelving unit as part of your coursework the college will probably supply the wood at cost price. Obviously, you will use a lot of tools while at college and it is a good idea to start buying your own tools for use once you are qualified. In fact, if you check out the 'carpenters wanted' ads in your local press, you will see that many actually specify only carpenters with their own tools will be considered. For example 'Carpenters wanted: 1st and 2nd Fix, must have CSCS card [which you have to pay for – see below] and nailgun' and 'Carpenters required: must have 110v tools'. Most carpenters will need saws, chisels, hammers, screwdrivers and drills. If you eventually decide to become self-employed you will need either a work van or, at the least, an estate car in which to travel with your materials to worksites. You will also have to invest in power tools such as sanders.

WILL PEOPLE WANT TO SEE PROOF OF MY QUALIFICATIONS?

These days, in many cases, yes they will. In 1995, in a bid to cut down on accidents and to improve the quality of construction work, the industry introduced CSCS (Construction Skills Certification Scheme) cards. These act as proof of your competence and qualifications and now cover 220 construction occupations, including bench joinery and carpentry. There are now over one million CSCS cardholders in the UK and you will be denied access to many building sites if you don't have one. By 2010 all workers within the construction industry will have to hold CSCS cards. The card itself costs £25 and you have to sit a health and safety test before one is awarded, which costs £17.50. For more information visit www.cscs.uk.com.

DID YOU KNOW?

The cruck frame – where a pair of timbers are tied or joined together at the top and raised to form a roof shape, linked by a ridge pole to another pair of timbers at the other end – is the oldest, strongest and simplest method of construction. Many houses and barns built from the 15th to the 17th century used this method and they are now making a comeback because they last for ever and are very environmentally friendly.

WILL I BE ABLE TO USE MY SKILLS ABROAD?

Yes. Great Britain is not the only country with a shortage of skilled craftspeople, and many carpenters go abroad to work on contracts. If you are contemplating using your skills abroad, you may like to think about taking a GCSE in a foreign language.

WHAT CAN I EXPECT TO GET OUT OF THE INDUSTRY PERSONALLY?

A great sense of satisfaction and achievement as you can actually see the finished product of your labours. Talk to any carpenter and they will tell you the most satisfying thing about their job is the fact that at the end of the day they can actually see what they have created. Working with your hands can give you a powerful sense of achievement. Also, if you work on site or in a workshop, there is the camaraderie that grows when you and your workmates are all concentrating on a common goal. For the self-employed freelancer there is the opportunity to work in a variety of different locations and to meet a cross-section of the public. Finally, these are skills you will never lose, so you'll never have to employ a carpenter to hang doors, fit windows or put down decking in your own home – you'll be able to do it all yourself.

HOW WILL THE WIDER PUBLIC SEE ME?

As a valuable member of society, offering a professional skill. Most people are unaware of the vast variety of jobs carpenters undertake. They usually only become aware of what they do when they need one to come and work on their properties! Every homeowner who has ever needed new cupboards built, or sash windows fitted, knows the value of a well-trained, professional carpenter, as does anyone who has employed a cabinet-maker to build a special item of furniture. As so many people no longer have woodworking skills, those who do become a precious commodity. No wonder the modern carpenter is so very much in demand.

COULD I BE MY OWN BOSS EVENTUALLY?

Of course you could. Many carpenters run their own businesses. This is especially true of freelance carpenters and those who are subcontracted to do work for a larger firm. Even if the thought of running your own business, with all the paperwork that entails, doesn't appeal to you, there are great opportunities to progress from a craftsperson up to the level of site manager or construction manager as there is currently a shortage of people with the necessary expertise.

6

SAMANTHA JAY COMPTON

Case study 2

Once I've completed this course I'm hoping to get on to the cabinet-making side of carpentry.

THE FIRST-YEAR APPRENTICE

20-year-old Samantha has always been good at working with her hands and knew a job employing this talent was what she wanted to do. While at school she did a course working with resistant materials and did BTEC Engineering level 3, in which she had to construct a steam engine. Since then she has been on a lot of courses, including a multi-trade course, which involved painting, decorating, some carpentry, bricklaying and welding. She also completed a one-year plumbing course. She started as a first-year apprentice at the Building Construction College in Stratford in 2007 and is now doing her level 1 Carpentry as an apprentice with the Sandwood Construction Company.

'I really like this course because it's really creative and hands on and once I've completed this in three years' time I'm hoping to get on to the cabinet-making side of carpentry. In a sense it's a really good skill to have because you can use it for yourself and you can also go on to teach your skills to other people. The only thing I find frustrating is when I fudge things up, when things don't go right. There's also the splinters, but they are really only a minor annoyance.

'I found my apprenticeship through ConstructionSkills – they sent me a list of names and addresses of companies that were willing to take on apprentices, I went for an interview with Sandwood and they sent me here. Since we started I've made lots of joints and now I'm constructing a T joint. Although I sometimes find the work physically slightly hard, and it takes me some time to get my head around the theory, I'm definitely going to stick with it because I'm enjoying it and being the only girl in my group doesn't faze me because I've already been on a lot of courses where I've been the only girl.

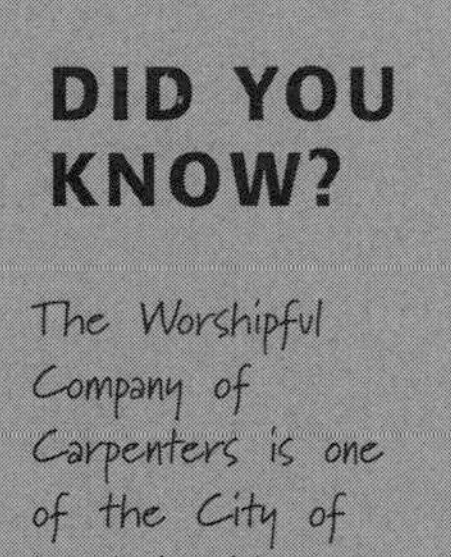

'In the long run I'm not really concerned about owning my own business, I'd rather be employed on a freelance basis by another company where I would get enough free time to do my own projects and where I could talk things over with colleagues, and if needed I'd be able to ask someone else for help.

The best thing about carpentry is the end result, because if you've done everything right you think 'Yeah, I made that!'

'The best thing about carpentry is the end result, because if you've done everything right and it comes out well you think "Yeah, I made that!" The reason I have done all these practical courses, including the plumbing, is because ultimately I want to be able to build my own house – from the foundations to the roof – and even furnish it with furniture I have made myself, so I'm definitely going to stick with it. This is really what I want to do.'

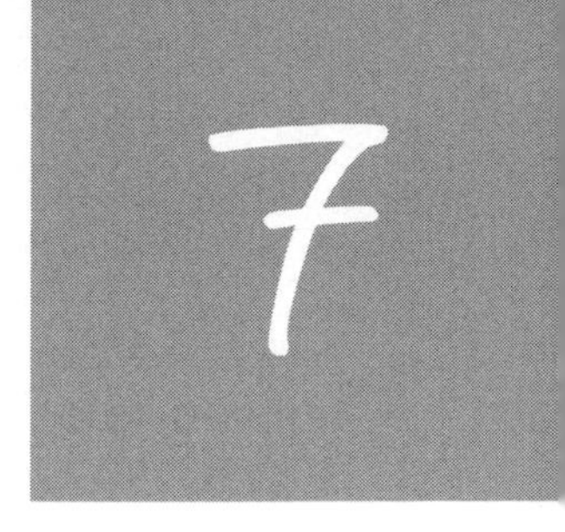

Training

By now, you should have a good idea if a job working with wood is for you. If you have decided it is, then you will need professional qualifications in order to advance as quickly and as far as you possibly can. Although there is no age limit for entry into the industry, most people start straight from school and this is because some training schemes do have an upper age limit. Carpentry has traditionally been a craft learned through an apprenticeship, in which a carpentry company or older, trained individual takes you on and teaches you the skills of the trade as you work. This is still the main way to get in to the industry and you can become an apprentice by approaching a company direct and becoming apprenticed to them or by joining a training programme such as the Construction Apprenticeship Scheme (CAS) run by ConstructionSkills (see below). You can also attend a college as a full-time student, or do a degree course at university. All these methods of training lead to qualifications recognised by the industry.

NATIONAL VOCATIONAL QUALIFICATIONS

National Vocational Qualifications (NVQs) or Scottish Vocational Qualifications (SVQs) allow you to learn practical skills on the job while training at college or a training centre, usually on a block-release or day-release basis. A two-year training programme will usually lead to NVQ/SVQ 2 with the option of completing a third year up to NVQ level 3.

An Apprenticeship (or Skillseekers in Scotland) is an extremely good way of learning your trade – so good that around 30,000 young people in the UK start some form of apprenticeship each year. It lets you earn as you learn and get qualifications. A Foundation Apprenticeship will lead to NVQ/SVQ level 2, while an Advanced Apprenticeship will lead to NVQ/SVQ level 3. You are eligible to do an Apprenticeship if you are between 16 and 24 years old. For more information on Apprenticeships visit www.apprenticeships.org.uk. In Scotland visit www.careers-scotland.org.uk, in Wales www.careerswales.com, and Northern Ireland www.delni.gov.uk.

CONSTRUCTION APPRENTICESHIPS SCHEME

In England and Wales a Construction Apprenticeship Scheme (CAS), run by ConstructionSkills, lasts for three years, while in Scotland there is a four-year apprenticeship registered with the Scottish Building Apprenticeship and Training Council. For both, entrants are registered with an employer and have an Apprenticeship Agreement that guarantees employment while they work towards NVQ/SVQ levels 2 or 3. You must be between 16 and 25 to be eligible for a CAS. The CAS complements the Apprenticeship scheme both in England and in Wales. For more information call the Construction Apprenticeship Scheme hotline on 0870 417 7274.

CITY & GUILDS

City & Guilds specialises in vocational qualifications and training, which means you undertake a course of learning while you are working. It has everything from National Vocational Qualifications (NVQ) and Scottish Vocational Qualifications (SVQ) to on-the-job training and Single Subject Diplomas. You can take City & Guilds certificates in Basic Carpentry and Joinery Skills, and Woodworking (Carpentry and Joinery). City & Guilds 'own brand' qualifications come in five levels, each level being roughly equivalent to the same level of NVQ or SVQ, although levels 4 and 5 are aimed at people who want to start their own businesses or go into management (such as the City & Guilds Higher Professional Diploma in Creative Arts, which would suit furniture designers who want to start their own companies). It also has a number of Design and Craft courses that are related to woodworking and carpentry, including chair caning, marquetry, woodturning, upholstery, staining and French polishing, furniture frame making, veneering and wood carving. These craft courses are offered at the following levels:

Design and Craft 7722 VRQ 1

This 30-hour introductory qualification provides you with the practical skills and knowledge needed to make a basic craft item or set of samples in your chosen subject. This qualification leads on to the level 2 Certificate in Design and Craft (7822/3).

Design and Craft 7822/3 VRQ 2

This qualification develops your craft skills to a high and saleable standard. There are two units you must complete in order to achieve this qualification. Unit 1 is the compulsory design unit, which helps students understand the way design ideas develop. Unit 2 is the specialist craft unit for the subject you have chosen to study (eg woodturning) which will help you gain practical skills and specialist knowledge. In order to complete the qualification you need to create three innovative artefacts. This qualification leads on to the level 3 Certificate and Diploma in Design and Craft (7922).

Design and Craft 7922 VRQ 3
If you want to develop your craft skills to a professional standard then you really should train to this level. For VRQ 3 you must make six items that not only demonstrate your design skills but must also be finished to a very exacting standard. If successful, you could move on to the Diploma in Design and Craft (7923).

Design and Craft 7923 Diploma VRQ 3
The level 3 Diploma is an advanced qualification for talented individuals who want to aim higher. Students research, plan and execute a project that allows them to explore and display their innovative designs and enhanced craft skills.

Upon completion of VRQ 3 you could progress on to level 4 or higher education.

For more information on the Design and Craft courses go to www.cityandguilds.com/designandcraft or call 020 7294 2800.

City & Guilds also offers a two-year, full-time Diploma course in Mastercrafts Diploma in Building Conservation and New Work (consisting of six areas – plastering, carpentry and joinery, sheet and cast metalwork, surface decoration, masonry and brickwork).

SCOTTISH QUALIFICATIONS AGENCY

For those wishing to become cabinet-makers, the Scottish Qualifications Authority (SQA) offers a Higher National Certificate (HNC) and a Higher National Diploma (HND) in Furniture Construction and Furniture Design.

BTEC

BTEC offers both National and Higher National awards such as National Certificate (NC) and National Diploma (ND), and Higher National Certificate (HNC) and Higher National Diploma (HND) in Furniture Design and Making, Furniture Design and Furniture Studies, which are very useful for those wishing to become cabinet-makers.

FT2 – FILM AND TELEVISION FREELANCE TRAINING

If you'd like to work in the glamorous world of showbiz, this could be right up your street. The FT2 scheme offers a two-year construction apprenticeship for people wishing to work on film and television sets. Obviously, the skills of a carpenter are essential for building sets and there is always plenty of work in England's busy studios such as Elstree and Pinewood. You have to be 16 or over to qualify for the scheme and will probably intersperse block release from college with actual work, eventually gaining an NVQ 3 qualification. (For more details see the Resources chapter, page 68.)

Most carpenters and cabinet-makers are trained to NVQ/SVQ levels 2 and 3, but if you want to gain promotion to a management position, such as site manager or contract manager, you can train up to NVQ/SVQ levels 4 and 5 and if you wish to take your education even further you can do a foundation degree (see below). Once you have a job in the industry, you can take a number of short courses that cover specific areas, such as the one-day Construction-Skills-approved Wood Machines Safety Course designed for machine operators. Your employer will usually pay for this type of course.

For most of the courses above there are no set entry qualifications, although it is good to be able to show a certain level of academic achievement. Good GCSEs to have are Maths, English, Science, and a craft or practical subject such as Woodwork or Needlework.

However, there are set entry qualifications both for the SQA courses and for the BTEC courses. For the SQA awards you will need four GCSE passes, grades A to C, with at least one A pass. For BTEC you will need four GCSE passes, grades A to C.

14–19 DIPLOMA

Currently being piloted is the new 14–19 Diploma in Construction and the Built Environment, which should go 'live' nationally in 2008. It is hoped that students who have taken part in the pilot scheme will join the workforce in time for the Olympic Games. There are two core compulsory learning units, 'The built environment and the construction industry' and 'Design for sustainability and the built environment'. There are then an extra seven technical, professional and craft units:

- Construction processes and technology
- Building design
- Surveying
- Carpentry and joinery
- Painting and decorating
- Brickwork
- Building services.

For a single award you must take one of these optional units and for a double award you must take three.

If you are interested in design, then the new Diploma in Manufacturing and Design (which is meant to go 'live' in 2009) could be more suitable.

WELSH BACCALAUREATE/SKILLS FOR WORK/OCCUPATIONAL STUDIES

The Welsh Baccalaureate (Welsh BAC) recognises almost anything young people do at school or college through a Core programme of activities, consisting of four components (Key Skills; Wales, Europe and the World; Work-related Education; Personal and Social Education) plus options such as GCSE, VGCSE, NVQ or BTEC. Go to www.wbq.org.uk for more information.

In Scotland, the new Skills for Work programme Intermediate 2 is a short course to give people real employability skills. The same is true for the Occupational Studies (entry level and levels 1 and 2) in Northern Ireland. These are like tasters so you can see if construction is really the career for you. Both these courses are roughly equivalent to the Young Apprenticeship in England. For more information on Skills for Work go to www.ltscotland.org.uk and for Occupational Skills www.deni.gov.uk.

YOUNG APPRENTICESHIPS

Young Apprenticeships combine traditional education and vocational education and give pupils work experience in schools so they learn employability skills such as time-keeping, communication and transactional analysis. The Young Apprenticeship in Construction was introduced in 2006.

Young Apprenticeships are for Key Stage 4 pupils who show real motivation and ability. They learn not only in the classroom, but also in colleges with training providers, and in the workplace. Pupils follow the core National Curriculum subjects and for two days a week they also work towards nationally recognised vocational qualifications (NVQs) in construction. For more information on Young Apprenticeships go to www.vocationallearning.org.uk/youngapprenticeships.

FOUNDATION DEGREES

These are primarily for candidates who are already in work. Delivered in partnership with employers, they are the training programmes for those who really want to progress to become supervisors, junior managers or to go on and own a business and can be studied part-time so trainees can continue working while they are learning. FDs are higher education qualifications set at one level below an honours degree. There are over 1,600 different FDs to choose from at present (with another 800 planned) and although there are no carpentry-specific courses you can do such topics as Building Renovation and Design, Building Crafts, and Sustainable Resources for the Built Environment. For more information go to www.foundationdegree.org.uk

CONSTRUCTION AWARD SCHEME/COLLEGE-BASED

If you don't want to go down the apprenticeship route (or if you can't find an apprenticeship) you can now also do a Construction Award, which is totally college-based and does not entail going out to work on site. However, if you combine your college training with work experience it can lead to an NVQ. The awards are at three levels: Foundation, Intermediate and Advanced and are aimed at pupils in Key Stages 1 to 3 and provide units of work to facilitate and support curricular activities such as Design and Technology,

English, Science, Mathematics and ICT. Students taking part in the scheme are expected to develop appropriate skills and knowledge associated with specific areas of the construction industry and to become more aware of the wider issues affecting planning and construction of commercial and domestic buildings. They should also become more aware of the effects of construction on the social, economic and environmental development of communities. The awards cover five main areas:

- Shelter
- Structure
- Materials
- Systems
- Careers.

For more information go to www.constructionawards.co.uk

BOAT BUILDING

If the building of beautiful, hand-finished boats appeals to you there is now a City & Guilds level 3 course in Boat Building, Maintenance and Support. The course takes place at the Boat Building Academy in Lyme Regis and lasts for 38 weeks. Only 16 students at a time can participate as it is very intensive, with a high teacher-to-student ratio. Units include:

- Lofting
- Traditional boat building
- Modern wooden boat building
- Fibre reinforced plastic boat building
- Yacht joinery
- Spar and oar making
- Finishing coatings and compounds.

For more information on the course and entry requirements go to the website at www.boatbuildingacademy.com or call 01297 445545.

DID YOU KNOW?

The Doorway is a loose alliance of (among others) the Timber Trade Federation, the Institute of Carpenters, the UK Timber Frame Association, the Carpenters' Company and the North West Timber Trade Association. It exists to promote careers in the wood trades – from forestry workers right through to joiners and designers. Its website has masses of information on jobs, qualifications and a live news feed. Go to www.thedoorway.org.uk

Getting a good academic record isn't the only positive thing you can do while at school to help your entry into the profession. One of the best things you can do is to get some work experience. As so many companies are eager to get the best trainees, they now offer work experience as a 'taster', letting you see what they actually do, and maybe encouraging you to do your apprenticeship with them. Talk to your careers officer or careers advisor about work experience opportunities in your area. Also ask them if your school is due to be visited by ConstructionSkills. Each October ConstructionSkills organises National Construction Week, which includes classroom visits by people who actually work in the industry.

If you have the facilities, actually building something is an excellent way to show off your skills. It need not be anything particularly complicated or big, but if it is well made and well finished it will show you have a talent for carpentry. If you are doing woodwork at school, this will be easy for you. If this is not practical, you can always show your interest in the subject by keeping abreast of new ideas, design trends and industry news by reading one of the many magazines dedicated to carpentry and cabinet-making. Details of the major publications are listed in the Resources chapter, page 68.

Whichever way you decide to train, finding the best training college for your chosen career is crucial. Many colleges have now become Centres of Vocational Excellence, which means they specialise in particular subjects. For instance, Lambeth College is the CoVE for the construction industry in London. Ask your careers officer if there is one in your area. You should also ask about the National Construction College. This is a network of colleges around the UK dedicated to training skilled construction workers (see the Resources chapter). Also find out which courses are actually available in your area. Check to see what would be most suitable for you by looking at the websites of the main awarders of vocational qualifications. All these addresses are included in the Resources chapter.

The guide on page 57 neatly sums up the various routes into a career in carpentry and cabinet-making, from the time you leave school, right up to the highest levels.

Well trained

What you learn as a trainee will very much depend on what branch of carpentry you decide to go in to. For instance, a cabinet-maker won't be up on a roof fitting beams, while a shopfitter won't need to know how to do marquetry. However, there are some aspects of working with wood that all trainees need to know about. These include:

- **Product recognition**
  One of the most basic skills anyone working with wood needs is how to distinguish between different woods; their particular properties; what they are used for; and how to preserve them. For instance, hardwoods can be very expensive and so are used only for certain projects. Other woods, such as walnut, are used for their decorative qualities. At college you will be taught how to recognise a great variety of woods and how they can be preserved.

- **Use of tools**
  Knowing which tools to use for which job is an essential skill. In a workshop there is a huge range of tools, from saws and screwdrivers to different hammers, chisels and planers. Power tools have become increasingly essential to all types of carpenters, not just those in workshops. You will

be taught how to use them effectively and, above all, safely.

- **Health and safety**
  Health and safety is of the utmost importance in carpentry. Not only do you need to know how to use your tools safely, but also to be aware of safety issues in your environment including wood dust, certain chemicals such as adhesives, paints and protective chemicals, and structural security. You will be taught the proper use of safety equipment such as facemasks and earplugs. It's not only your own safety that is important: you will also be taught how to work in a way that does not endanger the safety of others.

- **Fixings and hardware**
  Carpentry isn't just about wood, it is also about the fixings, such as screws and nails, used to hold items together or attach them to other things, and it is also about hinges, plates, brackets, locks, and handles. You will need to know what each different piece of hardware does and which pieces you need for each different job.

- **Basic craft skills**
  As a trainee you will be required to do a number of set-piece assessment projects where you actually make things. These will test your ability to do geometry, measure correctly, and put a wooden structure together properly. At NVQ level 2 projects

usually include putting together a full-size door and door frame, as well as making a cabinet of some description. As your training advances, so do the projects, to a level where you may have to complete a box sash window or a model staircase. Projects will vary depending on which course you are taking. For instance, a City & Guilds Diploma in Fine Woodwork includes practical experience with woodturning, hand veneering, marquetry and French polishing, while students doing an apprenticeship in shopfitting must have practical knowledge of machining from sawn stock, making a curved bench seating unit, and finishing complex internal panelling.

- **Business skills**
  Once again, depending on which course you are taking, you may well have to learn business skills such as estimating quantities, producing cutting lists, and measuring up and producing scale drawings based on a client's brief. Your course may also include key skills in numeracy, literacy, and information technology, all of which will be essential to you, especially if you decide to go freelance and set up your own business.

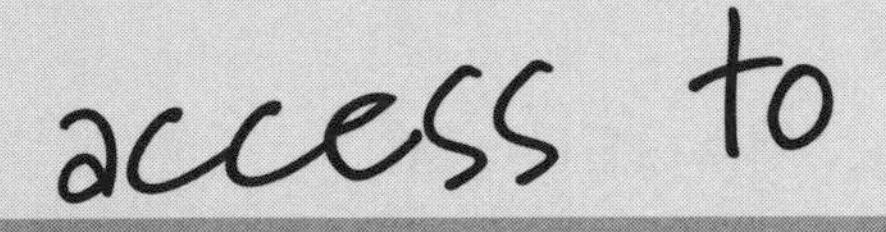

CARPENTRY AND CABINET-MAKING

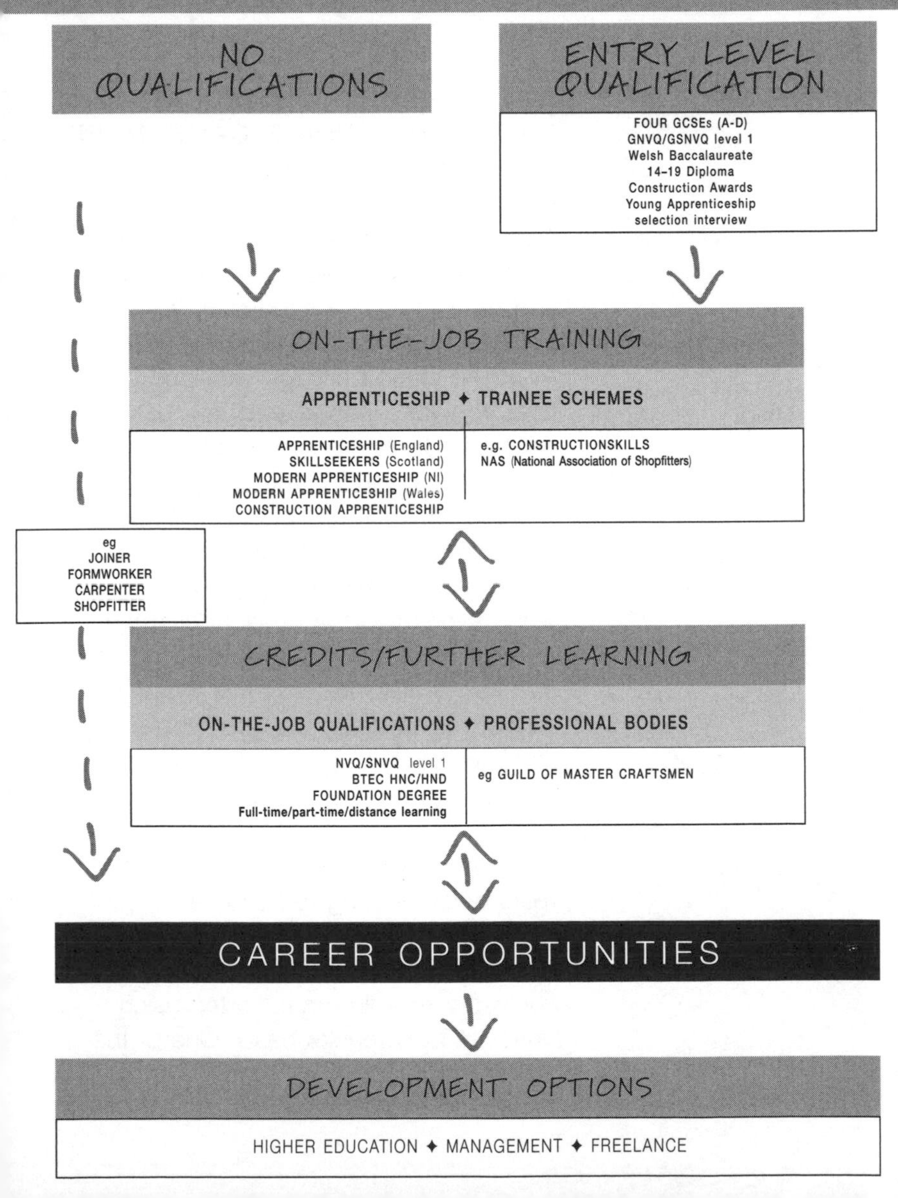

I like the creating and the satisfaction at the end of the day when you can see what you have built.

PAUL MAGERUM

Case study 3

THE MAINTENANCE CARPENTER
At 47 years old, Paul Magerum is the Maintenance Carpenter for Writtle Agricultural College in Chelmsford. As a boy he was interested in woodwork and he wasn't very academic, so he decided to go into the building trade. At 15 he started a three-year apprenticeship with a bespoke joiners based in Stratford. He worked on the job and went to college one day and two evenings a week. He started off doing menial jobs for the firm, but at the end of his apprenticeship he was doing everything the other tradesmen were doing. He particularly liked going out from the workshop to fix whatever had been made on site in places such as University College London, office blocks and in private homes.

Paul stayed with the firm for 10 years before leaving to become self-employed. He mainly worked on refurbishing buildings and became involved with a firm that started using him on a fairly regular basis. One of the firm's clients was the merchant bank Kleinwort Benson and he worked on their buildings as a maintenance carpenter for the next seven years. He particularly enjoyed working on the older buildings where he had to fix and replace wood panelling and oak doors. For the last two years at Benson he was the foreman. He then worked for another banking firm, Merrill Lynch, initially with responsibility for one building, but after seven years he was leading a large team and looking after 13 buildings.

Realising this was not what he wanted to do, he left to become self-employed again before getting the position at the agricultural college, which is on a 500-acre farm. Paul is responsible for the upkeep of 80 buildings, from sheds to office blocks. Last year, he built a wooden bridge from scratch and now everyone at the college uses it daily.

There are so many different aspects to carpentry you really need to be a good all-rounder to make a go of it.

'When I left my first employers to go self-employed I suddenly started earning about three times the amount of money I earned before and it is good to bear in mind that workshop-based carpenters and joiners are probably paid the least in the industry. Initially, I enjoyed working with a big team at Kleinwort Benson, but towards the end of my time with Merrill Lynch I was off the tools and basically doing a desk job and I slowly began to realise it wasn't me. It was slowly killing me and so I reassessed my life. I realised I missed the hands-on stuff and I wasn't enjoying what I was doing. Now I'm back using tools again, using my hands, and as there are so many buildings here the work is very varied. One day I'll be putting up a partition wall, or working on a barn, or even building a roof.

'I like the creating and the satisfaction at the end of the day when you can see what you have built. I really like working with my hands and with tools and the good thing the Merrill Lynch job taught me was that I am not a boss or a manager. I like working by myself on projects, I prefer it, so that is one aspect of this job that really suits me. Plus, it is only five minutes from where I live so I don't have to commute like I did when I worked in the City, and also, I am out in the fresh air all day.

'Communication is definitely one of the skills you need for this job. Because of my experience I can communicate with everyone from the labourer right up to the managing director. You also have to really enjoy what you are doing. There are so many different aspects to carpentry you really need to be a good all-rounder to make a go of it. There is a massive, massive shortage and you can make a really good living from it, but you've just got to really want to do it. For me it is a passion, like being a chef, and I'm enjoying what I am doing now more than I ever have and that's saying a lot considering I started at 15.

'It was a fantastic feeling when I finished that bridge; a year later, I gave it a coat of linseed oil and everyone who walks over it comments on it. How many people get a chance to do stuff like that? Knowing what I know now, if I was starting out in carpentry again I would concentrate on working with traditional tools on traditional buildings because that is what I am interested in. For instance, if I had an opportunity to work on the reconstructed Shakespeare's Globe theatre, well that would just be a dream come true for me. There are those fantastic jobs out there, and that's what I'd do.'

9 Career opportunities

Earlier, in the What's the story? chapter, we looked at just some of the jobs open to trained carpenters. Even before you start to train it is a good idea to look at what job opportunities could be there for you in the future. The skills you learn as you train will stay with you for the rest of your life and they can take you right to the top of your profession. Take the career path of Richard Easton, for example. He started off at 16 as an Apprentice Joiner doing his City and Guilds at A E Hadley Ltd. He was then promoted to Setter Out, became an Estimator with the firm and rose to Director of Sales and Overseeing Estimating. After a spell as a Co-Director he went on to become Managing Director of the whole firm. Your ambitions may not be this high, but there will be plenty of different career paths for you to take. The following diagram will give you a rough idea of just how far you can go.

We have already discussed what formworkers, joiners, shopfitters, and cabinet-makers actually do, but as you advance up the career ladder, your role and your responsibilities will obviously change. For a start, you may find your job becomes less 'hands on' and more involved with paperwork and co-ordination. For example, **setters out** are very skilled shopfitters who prepare working, full-size or

This is an industry with real opportunities for advancement. As you go through the training and discover where your strengths lie you will be able to map out a future career path. The diagram below shows options that will open up to you once you have trained.

CAREER OPPORTUNITIES

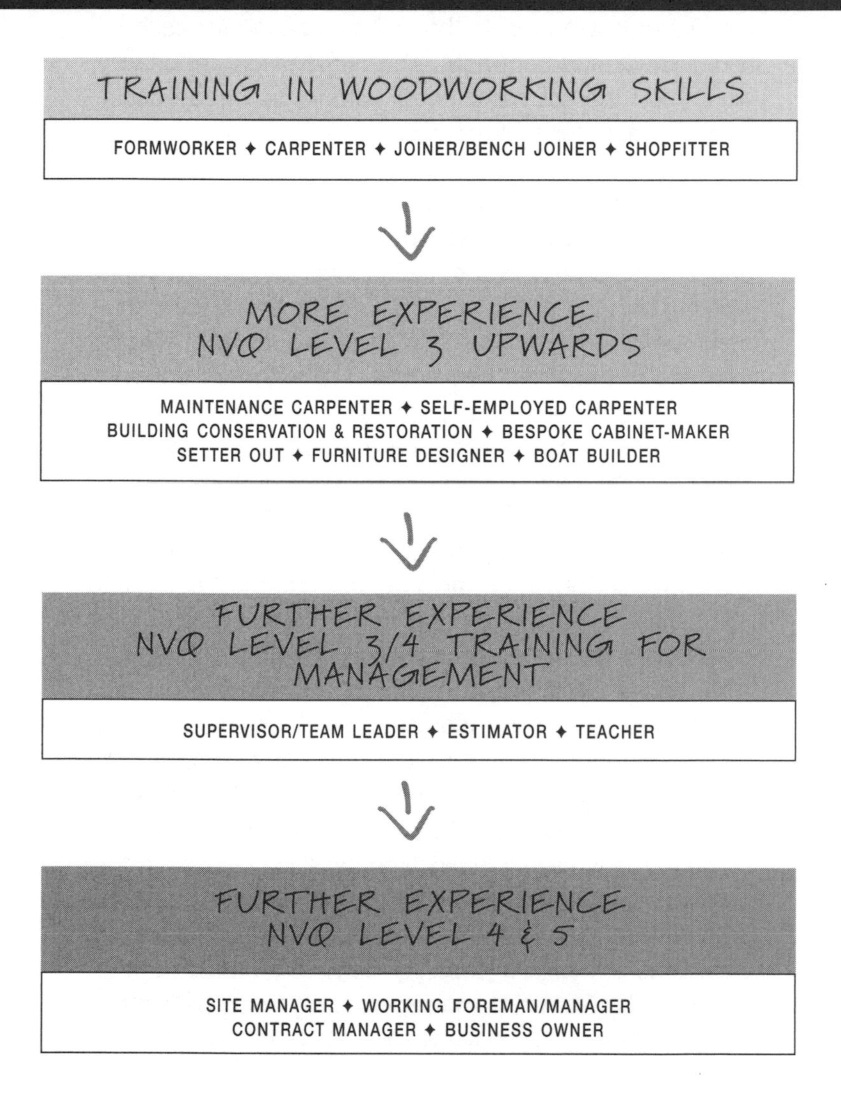

scale drawings of the fittings to be made, while the actual cutting and fitting together of the cabinets or shelving units is carried out by someone else. **Estimators** work out the amount of raw materials that will be needed for each job and price up the contracts. This is extremely important work, for if the estimator gets it wrong, he or she could lose contracts for their company.

When you get to management level you may find you do no actual woodwork at all, but spend your days on the phone to suppliers, co-ordinating jobs with other people working on your site such as plumbers and electricians, communicating with your staff, and keeping your clients informed of progress. You will also have to sort out any problems that may arise, such as a job going seriously over-budget. **Supervisors**, **site managers** and **team leaders** need exceptionally good people skills as they oversee other people's work on site and have to offer advice, make suggestions and sort things out when a job goes wrong. They must also be very aware of the health and safety aspects of the job, for example ensuring workers wear safety gear such as steel-toed boots and helmets while on site. **Working foremen** or **managers** are responsible for running workshops and looking after the joiners who work there. Quality control is a big part of their job and so they have to pay attention to detail. They also have to ensure the finished product is made and delivered on time: many clients insert time-sensitive clauses in their contracts, which means that if their order is not delivered on time they pay less. Once again, they must be very conscious of health and safety, as joiners use a lot of power tools. **Contract managers** have many different responsibilities. They have a supervisory role on site and oversee the whole project, from the planning stage right through to the final product. They have to make sure that even the smallest details in the contract are not overlooked and will liaise closely with the client.

As you progress up the promotion ladder, you will dramatically increase your earning potential, so it makes sense to get the best training you can. Remember, a carpenter's skills are highly regarded and their services much sought after, so if this is the career you have chosen to pursue, make sure you take advantage of all the opportunities to learn that are available to you.

10 The last word

By now you should have a pretty good idea what working as a carpenter or a cabinet-maker actually entails. Mankind has been working with wood since the dawn of civilisation and even with the advent of man-made materials such as plastic and concrete the skills of the woodworker are still very much in demand. In fact, as we become more environmentally friendly, owning things made from wood and conserving them so they last are becoming increasingly important. Deciding on a career in carpentry will open up a world of opportunities for you. Overleaf is a fun checklist to see if carpentry really is the career choice for you.

There has never been a better time to train to become a carpenter. With a shortage of skilled craftsmen across the whole industry, and an upsurge in major construction projects across the country, well-trained, professional carpenters are now at a premium. This is varied, interesting, and above all fulfilling work where you can actually see the fruits of your labours. It can also be very rewarding financially, with very talented craftsmen earning substantial amounts for their expertise. There are also plenty of opportunities to advance to management level and beyond. The future for the whole industry is looking bright and hopefully this book has helped you to make up your mind whether or not you want to be part of that industry. The following chapter contains comprehensive information on the organisations you should contact if you want to make a career in carpentry.

If you have made it this far through the book then you should know if **Carpentry and Cabinet-Making** really is the career for you. But, before contacting the professional bodies listed in the next chapter, here's a final, fun checklist to show if you have chosen wisely.

THE LAST WORD

✔ TICK YES OR NO

DO YOU LIKE WORKING WITH YOUR HANDS? ☐ YES ☐ NO

DO YOU CONSIDER YOURSELF CREATIVE? ☐ YES ☐ NO

DO YOU WANT A JOB WHERE YOU WILL BE DOING SOMETHING DIFFERENT EVERY DAY? ☐ YES ☐ NO

ARE YOU SELF-MOTIVATED AND ABLE TO THINK ON YOUR FEET? ☐ YES ☐ NO

ARE YOU ABLE TO COMMUNICATE EFFECTIVELY WITH PEOPLE? ☐ YES ☐ NO

ARE YOU A SELF-STARTER, ABLE TO TAKE CONTROL AND RESPONSIBILITY? ☐ YES ☐ NO

If you answered 'YES' to all these questions then CONGRATULATIONS! YOU'VE CHOSEN THE RIGHT CAREER!

If you answered 'NO' to any of these questions then this may not be the career for you. However, there are still some options open to you, for example you could work as a Setter Out or a DIY Shop Assistant

11 Resources

In this section you will find all the addresses, telephone numbers and websites for the relevant government and industry advisory and training bodies for the construction industry. There is also a list of publications you may find useful to read.

TRAINING AND ADVICE

CAREER DEVELOPMENT LOANS

Packs available from 0800 585505
www.direct.gov.uk/cdl

If you are undertaking a vocational training course lasting up to two years (with one year's practical work experience if it is part of the course) you may be eligible for a Career Development Loan. These are available for full-time, part-time and distance learning courses and applicants can be employed, self-employed, or unemployed. The government pays interest on the loan for the length of the course and up to one month afterwards.

CITY & GUILDS

1 Giltspur Street
London EC1A 9DD
020 7294 2800
www.cityandguilds.com

City and Guilds is the leading provider of vocational qualifications in the United Kingdom. It has five different levels of qualification, with 1 based on the lowest competence level, and 5 based on the highest, and it offers

NVQ and SVQ to Apprenticeships and Higher Level Qualifications. The excellent website lists all the qualifications it provides in carpentry and cabinet-making (search under Construction).

CONNEXIONS
www.connexions.org.com

The Connexions service has been set up especially for 13- to 19-year-olds and offers advice, support and practical help on many subjects, including your future career options. In the Career Zone on the site you will find Career Bank, offering information on training and jobs in carpentry and cabinet-making.

CITB-CONSTRUCTIONSKILLS
Bircham Newton
King's Lynn
Norfolk PE31 6RH
01485 577577
www.citb-ficonstructionskills.co.uk

ConstructionSkills runs most of the employer-based training in the UK and is committed to training professionals to a very high level. If you wish to get a place on its Construction Apprenticeship Scheme you can apply to ConstructionSkills direct. It has some very good pamphlets and brochures offering more information, including a sheet entitled Wood Occupations and brochures on Shopfitting and on Building Conservation and Restoration. Alternatively have a look at the website.

In Scotland:
4 Edison Street
Hillington
Glasgow G52 4XN
0141 810 3044

In Wales:
Units 4 and 5, Bridgend Business Centre
David Street
Bridgend Industrial Estate
Bridgend CF31 3SH
01656 655226

EDEXCEL
190 High Holborn
London WC1 7BB
0870 240 9800
www.edexcel.org.uk

Edexcel has taken over from BTEC in offering BTEC qualifications, including BTEC First Diplomas, BTEC National Diplomas, and BTEC Higher Nationals (HNC and HND). It also offers NVQ qualifications. The website includes qualification 'quick links' and you can search by the qualification or the career you are interested in.

FEDERATION OF MASTER BUILDERS
14–15 Great James Street
London WC1N 3DP
020 7242 7583
www.fmb.org.uk

This is the largest trade association for the UK construction industry and represents over 13,000 small and medium-sized building companies.

FT2 – FILM AND TELEVISION FREELANCE TRAINING
www.ft2.org.uk

Offers two-year construction apprenticeships for people aged 16 and over who want to work on film and TV sets. Click on Training for information on its carpentry apprenticeships.

GUILD OF MASTER CRAFTSMEN
166 High Street
Lewes
East Sussex BN7 1XU
01273 477374
www.guildmc.com

Thousands of companies up and down the country belong to the Guild, which endeavours to maintain and uphold standards of excellence within the industry.

INSTITUTE OF CARPENTERS (CENTRAL OFFICE)
35 Hayworth Road
Sandiacre
Nottingham NG10 5LL
0115 949 0641
www.carpenters-institute.org

The IOC is a craft association for everybody working in the wood trades.

LEARNING AND SKILLS COUNCIL
Apprenticeship helpline
0800 015 0600
www.lsc.gov.uk
www.realworkrealpay.co.uk

Launched in 2001, Learning and Skills Council now has a main office in Coventry and nine regional offices. It is responsible for the largest investment in post-16 education and training in England and this includes further education colleges, work-based training and workforce developments. For MAs in Scotland you should look at www.modernapprenticeships.com or www.careers.scotland.org.uk.
In Wales you should look at www.beskilled.net.

NATIONAL ASSOCIATION OF SHOPFITTERS (NAS)
NAS House
411 Limpsfield Road
The Green
Warlingham
Surrey CR6 9HA
01883 624961
www.shopfitters.org

NAS welcomes applications for training from all eligible young people, regardless of sex, race or disability. It runs its own annual awards competitions, including the Second Year Apprentice Award.

NATIONAL CONSTRUCTION COLLEGE (NCC)
0870 4166 222
direct.training@citb.co.uk

The NCC is a network of five colleges around the country that specialise in training young people as skilled operatives and potential supervisors in the construction industry. All courses combine a mixture of site work and residential training and last between 4 and 43 weeks.

NEW DEAL
0845 606 2626
www.newdeal.co.uk

If you are an older individual looking to change careers and you have been unemployed for six months or more (or receiving Jobseekers Allowance), you may be able to gain access to NVQ/SVQ courses through the New Deal Programme. People with disabilities, ex-offenders and lone parents are eligible before reaching six months of unemployment. Check out the website for more information.

QUALIFICATIONS AND CURRICULUM AUTHORITY (QCA)
83 Piccadilly
London W1J 8QA
020 7509 5555
www.qca.org.uk

In Scotland:
Scottish Qualifications Authority (SQA)
Optima Building
58 Robertson Street
Glasgow G2 8DQ
Customer Contact Centre: 0141 242 2214
www.sqa.org.uk

These official awarding bodies will be able to tell you whether the course you choose leads to a nationally approved qualification such as NVQ or SVQ.

WORSHIPFUL COMPANY OF CARPENTERS (THE CARPENTERS' COMPANY)
Carpenters' Hall
Throgmorton Avenue
London EC2N 2JJ
020 7588 7001
www.thecarpenterscompany.co.uk

One of the City of London's oldest livery companies, The Carpenter's Company was founded in 1271. It runs the Building Crafts College and provides scholarships in the trade as well as sponsoring craft competitions.

Building Crafts College
Kennard Road
Stratford
London E15 1AH
020 8522 1705

PERIODICALS

Cabinet Maker
7th floor
Ludgate House
245 Blackfriars Road
London SE1 9UR
020 7921 8406
www.cabinet-maker.co.uk

This weekly publication is one of the best-known titles for the industry.

Furniture and Cabinet Making
86 High Street
Lewes BN7 1XN
01273 477374
www.thegmcgroup.com

The monthly magazine of the Guild of Master Craftsmen: aimed at makers of bespoke fine furniture. The GMC Group also publishes *Woodcarving* and *Woodturning*.

Woodworking News
Old Sun
Crete Hall Road
Northfleet DA11 9AA
01474 536535
www.nelton.co.uk

This magazine is published ten times a year and is aimed at craftsmen, giving product and industry news. Nelton, the publishing company, has two other titles that may be of interest: *Irish Woodworking and Furniture News*; and *Furniture Products*.